How Does It Grow?

# HOW DO SUNFLOWERS GROW?

By Kathleen Connors

Gareth Stevens
PUBLISHING

**Please visit our website, www.garethstevens.com. For a free color catalog of all our high-quality books, call toll free 1-800-542-2595 or fax 1-877-542-2596.**

**Library of Congress Cataloging-in-Publication Data**
Names: Connors, Kathleen, author.
Title: How do sunflowers grow? / Kathleen Connors.
Description: New York : Gareth Stevens Publishing, [2022] | Series: How does it grow? | Includes index.
Identifiers: LCCN 2020011763 | ISBN 9781538263112 (library binding) | ISBN 9781538263099 (paperback) | ISBN 9781538263105 (6 Pack)| ISBN 9781538263129 (ebook)
Subjects: LCSH: Sunflowers–Juvenile literature.
Classification: LCC SB299.S9 C66 2022 | DDC 583/.983–dc23
LC record available at https://lccn.loc.gov/2020011763

First Edition

Published in 2022 by
**Gareth Stevens Publishing**
111 East 14th Street, Suite 349
New York, NY 10003



Editor: Kristen Nelson
Designer: Katelyn E. Reynolds

Photo credits: Cover, p. 1 Pasuk Lertbuaban / EyeEm/Getty Images; p. 5 Mike Powles/Stone/Getty Images; pp. 7, 24 (seeds) MirageC/Moment/Getty Images; p. 9 Les Cunliffe/ Photolibrary / Getty Images Plus; p. 11 Elva Etienne/Moment/Getty Images; p. 13 EllenMoran/E+/Getty Images; pp. 15, 24 (sprout) Ed Reschke/Stone/Getty Images; p. 17 miguelangelortega/Moment/Getty Images; pp. 19, 24 (bud) Kanchanalak Chanthaphun / EyeEm/Getty Images; p. 21 Geri Lavrov/Moment/Getty Images; p. 23 Sean Seltzer / EyeEm/Getty Images.



Printed in the United States of America



CPSIA compliance information: Batch #CSGS22: For further information contact Gareth Stevens, New York, New York at 1-800-542-2595.

# Contents

Sunflowers are pretty.
How do they grow?

Sunflowers grow
from seeds.

They are planted
in spring.
They grow in soil.

They can be in
the ground or a pot.

Sunflower seeds
need water.
They need lots of sun.

Tiny plants come out of seeds.
These are sprouts.

They pop out of the soil.
They grow leaves.

They can grow tall stems.
They grow buds.
Buds open as flowers.

Sunflowers are yellow. They can be orange or red too.

Their center is brown.
This is where seeds are.

# Words to Know

bud

seeds

sprout

# Index